The Unofficial Essential Skills / Revision Guide for MFM1P

Grade 9 Applied Mathematics in Ontario

By Mark Burke

ISBN : 978-1479372713

How to use this booklet

This booklet is designed to cover the basics of the grade 9 academic mathematics module MFM1P. It is aimed primarily at students who are anxious about obtaining a Level 3. Students who are aiming for level 4 will find this a useful resource for revision but should aim to stretch themselves with more extension material. *Nonetheless , this booklet does cover all of the specific expectations as outlined by the Ministry of Education document 'The Ontario Curriculum Grades 9 and 10 : Mathematics (2005).*

Students should *work* through the examples rather than just read them, and attempt *every* question. Solutions and answers are at the end of the booklet.

Please note diagrams are not to scale.

For the purposes of simplicity and saving space some answers have been prematurely rounded which has affected their accuracy. Students should only round final answers, using previous answers stored in the calculator or at working with at least 5 decimal places.

Section A : Number Sense and Algebra

Operating with exponents

Substituting into and evaluating algebraic expressions involving exponents

- Calculating with exponents means multiplying a number by itself however many times the exponent number is ; e.g. 9.8^3 means 9.8 x 9.8 x 9.8 = 941.192

- Calculating fractional exponents by hand means multiplying the numerators(tops) by each other and the denominators (bottoms) each other, just as in fraction multiplication. e.g:

$(3/2)^3$ = (3/2) x (3/2) x (3/2)

= (3x3x3) / (2x2x2)

= 27/8

NOW YOU TRY 1 :

i) Without a calculator, evaluate $(5/2)^2$

ii) With a calculator, evaluate 7.2^4 to 1 d.p

- Describing areas and volumes in terms of algebraic expressions requires that you understand how powers of letters are added. e.g :A cube has width, length and height 'x '.

Volume = (X) (X) (X) = X^3. (although we don't write the 1's, imagine each x is to the power 1).

If there are numbers in front of the x's they are multiplied. e.g if a cuboid has dimensions length 2X , width 3X and height 4X, then

V = (2X) (3X) (4X) = $24X^3$

Surface area would use the fact that a cube has six faces, and the area of each face is X times X, eg. X^2

Surface area = (6)(X x X) = $6X^2$.

Note that if adding X's and there are numbers in front of the X , you just add those too.

Remember : 2X + 3X = 5X

BUT   2X x 3X = $6X^2$

Find and apply the exponent rules for multiplying and dividing monomials.

If you simplify (2 x 2) x (2 x 2 x 2) you see that it can be written as 2^2 x 2^3 = 2^5. This leads us to an important exponent rule.

Eg. $(6x) \times (5x^2) = 30x^3$ (*add* powers when *multiplying* bases)

$90x^3 / 20x^2 = 9x/2$. (*subtract* powers when *dividing* bases)

- Apply exponent rules to expressions involving two variables, e.g

$(3xy)(2x^2y) = 6x^3y^2$ (note how only powers of *like* bases are added)

NOW YOU TRY 2 :

i) A cuboid has length X, Width 2X and height 3X. Write expressions for the surface area and volume. (you should draw a diagram).

ii) Multiply : $(6x^2) \times (8x^3)$

iii) Divide : $50x^4 / 10x$

iv) Divide : $40x^3y^2 / 8xy^2$

Manipulating Expressions and Solving Equations

- Simplifying numerical expressions with integers and rational numbers, e.g.

$6 + 8 \times 2 - 3 = 6 + 16 - 3 = 19$ (you should know the laws of BEDMAS where *multiplication or division outranks addition and subtraction. Exponents outrank all)*

e.g 2 : $5 + 6 \times 8^2 = 5 + 6 \times 64$

$= 5 + 384$

$= 389$

Solving expressions involving percent, ratio, rate and proportion

<u>Percentages :</u>

To calculate a percentage of something, simply multiply it by the desired percentage number over 100. e.g

'A shirt is on sale for 80% of the ticketed price of 30 dollars. Calculate the new price' :

$30 \times (80/100) = 24$ dollars.

Or $30 \times 0.80 = 24$

To find percentage *change* , always calculate change as a percentage of the *original*. e.g :

A car was reduced in price from $20,000 to $17,000. What was the discount expressed as a percentage decrease ?

3000/ 20,000 = 0.15 = 15 %

- **Ratios** are like fractions that must be kept in balance. If one of the parts is multiplied by a number, the other parts must be multiplied by the same number. e.g :

- A cake is made using butter, sugar and flour in the proportions of 3 : 2 : 5. If the cake is made using 60 g of butter, how much sugar and flour are needed ?

60 / 3 = 20.

If we multiplied 3 by 20, we must do the same to all the other parts :

So the proportions are : 3x20 : 2 x 20 : 5 x20

= 60g : 40g : 100g, which is still in the same ratio as   3 : 2 : 5

Ratios can be simplified like fractions, e.g the ratio 2 : 4 reduces to 1 : 2 just as 2/4 simplifies to 1 : 2

Ratios can be used to calculate missing dimensions. E.g If a skateboard ramp has ratio of height to base of 2 : 3 then, what is the base if the height is 4.5 m ?

The missing base can be found like equivalent fractions with a multiplicative ratio. Compare the ratios 2 : 3 to 4.5 : X . If you wrote these as fractions, how would you get from 2 to 4.5 by multiplying ?

Divide the bigger number by the smaller number : 4.5 / 2 = 2.25. This means we are using the multiplier 2.25.

Apply this to the 3 then ; 3 x 2.25 = 6.75

So 2 : 3 = 4.5 : 6.75

Remember most ratio problems involve finding out how much one part is to be as a quantity. e.g $60 is to be split amongst 3 children in the ratio   2: 3 : 5 .   Adding the total parts give us 10. If $60 is split into 10 parts, each part is worth 6 dollars. So the split is :

(2 x6) : (3x6) : (5x6) = $12 : $ 18 : $ 30

<u>NOW YOU TRY 3:</u>

i) Simplify : -34 + 8 x 2

        ii)   A car has $1500 deducted from the sticker price of $23,000. How much as a percentage was deducted ?

iii) A 1200g chocolate bar is shared between 3 children in the ratio 1: 2 : 3. How much does each child get ?

Using inverse/reverse operations to simplify and solve equations :

Generally you reverse the laws of BEDMAS on a variable to find out its' value. Remember also the golden rule of equations : ' EQUATIONS MUST BE BALANCED ; WHAT YOU DO TO ONE SIDE YOU MUST DO TO THE OTHER '

Solving Equations involving Squaring and square rooting

Eg1. $5x^2 - 4 = 21$

$5x^2 = 21 + 4$ (Add 4 to both sides)

$5x^2 = 25$

$x^2 = 5$ (divide both sides by 5)

$x = +/-2.24$ (2.dp.)

e.g 2 (Squaring) : $\sqrt{(4X)} + 3 = 10$

It is a good idea to isolate the square root first.

$\sqrt{(4X)} = 7$

Now reverse the root by squaring :

$[\sqrt{(4X)}\,]^2 = 7^2$

$4X = 49$

$X = 49/4 = 12.25$

NOW YOU TRY 4 :

Solve for x : i) $3x^2 - 9 = 12$

ii) $\sqrt{2x} - 4 = 8$

Operations with monomials and polynomials.

The golden rule for operating with algebraic expressions is : 'only add *like* terms'. But what is a like term ? *A like term must contain the same letter, and to the same powers.*

i.e X's can only add with X's, and X^2 can only add with other X^2.
Eg :

$2X + 3y - X + X^2 + 4y - y^2 + 2X^2 = (2X^2 + X^2) + (-y^2) + (2X-X) + (3y + 4y)$

$= 3X^2 - y^2 + X + 7y$

Expanding with brackets

Every term inside the bracket must be multiplied by the term outside :

$2x(x+4) = 2x^2 + 2x(4) = 2x^2 + 8x$

Remember to add exponents where appropriate : eg.2

$2x^2(3x^2 - 2x + 1) = 6x^4 - 4x^3 + 2x^2$

Expanding *and* Simplifying :

After expanding, you must be careful to collect like terms and simplify if possible. E.g

$2x(4x+1) - 3x(x+2)$

$= 8x + 2 - 3x^2 - 6x$

$= -3x^2 + 2x + 2$

(note how x' simplified)

Remember again : there is one golden rule of manipulating all algebraic equations ; ' *What you do to one side of the equation you must also do the same to the other side*'
Most of the time this involves reversing operations to isolate a variable, and making sure you *apply the reverse operation to both sides*.

For example, re-arrange the equation below so that 'y'

is the subject (y = ...)

4x – 3y = 12

- 3y = 12 - 4x (subtract '4x' from both sides to start isolating 'y').

y = (12 – 4x)/ (- 3) (divide both sides by -3 to get one single, positive y)

y = 4 + (4/3)x (*every* part of the equation on the left is divided by 3)

NOW YOU TRY 5 :
 Rearrange so that 'r' is the subject of the formula ;
i) C = $2\pi r$ ii) A = πr^2

Solving first degree equations

Most basic equations involve these two basic steps ; using an additive inverse and then a multiplicative inverse.

This method can also apply to *solving* basic equations. You may also have to collect like terms together before applying the multiplicative inverse.

For example ; solve :   3x - 6 = 6x + 9
 Try collecting the x's on the left side, and constants (plain numbers) on the right :

3x − 6x = 9 + 6   (note how the numbers that moved changed signs +/-)
−3x = 15
x = 15/ −3 = − 5

Some of the trickier problems involve fractions which you can cancel by multiplying by the denominator of the fraction. For example, solve :

(1/3)x + 5 = 7

(1/3)x = 2

3 (1/3) x = 2 x 3   (cancel the 1/3 by multiplying by 3)
x = 6

<u>Using equations to solve problems</u>

Using X as a reference point for other unknowns, you can solve problems like :
' John is 8 years older than Mike, and and Mike is twice as old as Nada. The total of their ages is 83.'

Using Mike = X , John is then X + 8, and Nada is ½X
X + (X + 8) + (½X) = 83.
2½X = 83 − 8
X = 75 / 2½
X = 30
So Mike is 30, John is 38, and Nada is 15.

NOW YOU TRY 6 : Solve each equation for x :

i) $½x - 3 = 9$
ii) $5x - 3x = 6x + 10$
iii) $7x + 3 = 12 + 3x$
iv) $½x + 4x = 8 - ⅖x$

Section B : Linear Relations

The meaning of scatter-graphs

If you plot a set of data on an x-axis, and a corresponding set of data on a y-axis, you will get a scatter-plot which helps you to visualise the relationship between them.
Using the table of values below, plot the pairs as a co-ordinate on a graph.
e.g . The first point would be (6, 100)

X (age in years)	6	8	10	12	14
Y (height in cm)	100	110	120	130	140

Your graph should look something like the one below :

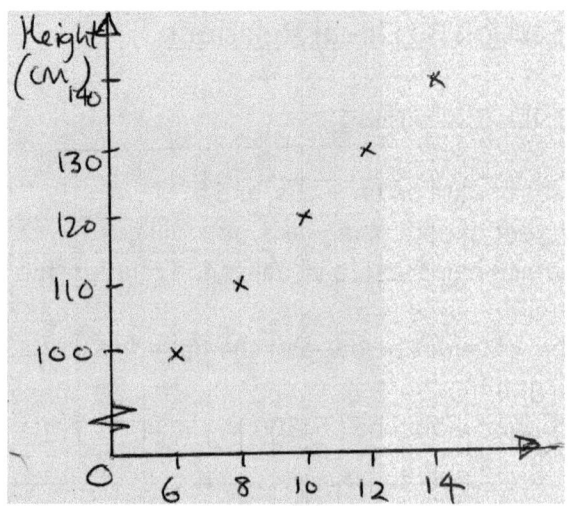

Whenever anything like a straight line is made on a scatter graph, you can assume there is some kind of relationship. You now need to know how to describe that relationship in mathematical terms :
We talk of trend meaning there is some sort of relationship. If the dots line up at all, even roughly, then there is a trend.

The strength of a trend can be called correlation. Strong correlation looks like a line (perfect correlation is a straight line) and weak correlation is where a rough, loose relationship occurs. In weak correlation the points are scattered but you can still see a pattern.

The type of trend can be positive or negative. If both quantities increase together (as they do here) then the trend is positive. If the trend line goes down as you read the graph left to right, then the trend is negative. In our first graph we have an example perfect, positive correlation. Below is an example of <u>weak, negative correlation.</u>

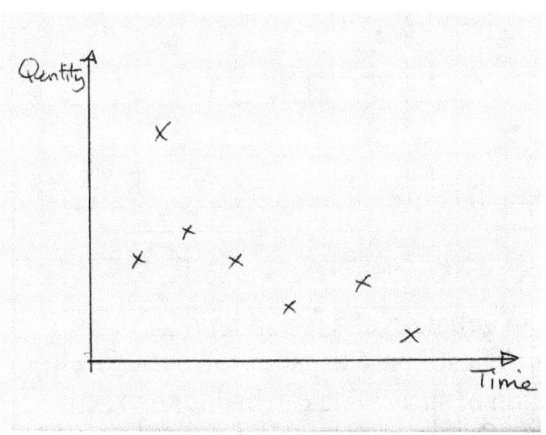

<u>Note that a line could be drawn roughly close to all points and this would be the line of best fit :</u>

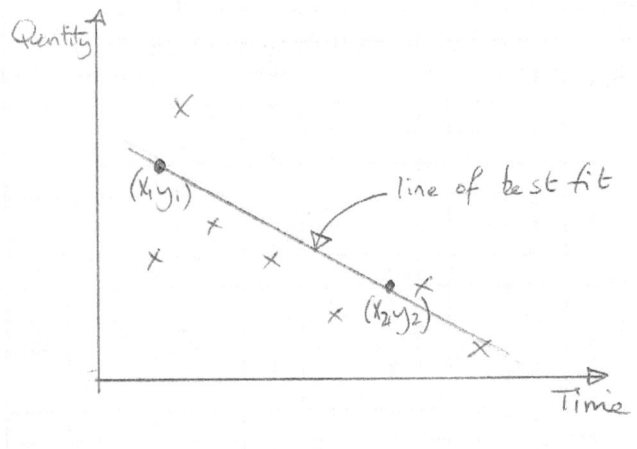

Determining the equation of a line of best fit.

Although lines of best fit are usually drawn roughly by eye, the equation of such a line can be found exactly. Pick two points on the line of best fit (Not from the original co-ordinates) and *calculate the slope as vertical change divided by horizontal change.*
You may use the formula $m = (y_2 - y_1)/(x_2 - x_1)$
Note that $(x_1\ y_1)$ *is always the point on the left*.

Example : a line of best fit is drawn on a scatter-graph and the points (4. 7) and (6, 8) are derived.
The slope is (8-7 / 6-4) = 1 / 2 .
The equation of a line is y = mx +b where m= slope and b is the line cuts the y-intercept.
 In this case y = 1/2x + b.
Use one of the points to substitute in x and y values and

find b :
If x = 4 when y =7, then : 7 = 1/2 (4) + b
7 = 2 + b
b = 5.
The equation of the line of best fit is ; y = ½ x + 5.
The equation of the line can used to roughly predict values on the graph.
For example, if x = 6, then y = ½ (6) + 5 = 8.

Interpolation and Extrapolation
When the equation of a line of best fit is restricted *within* the original range of values, the predictions of the equation are roughly reliable the process is *interpolation*.

X (age in years)	6	8	10	12	14
Y (height in cm)	100	110	120	130	140

For example, in our original example we can derive a line of best fit. Using (6, 100) and (8, 110) we derive the slope m = 10/2 = 5.
Using the point (6, 100) we have the equation :
100 = 5 (6) +b
b = 70
The equation of the line of best fit is y = 5x + 70.
Interpolation can be used to predict values between the

given range of x-values from 6 to 14.
If x = 11, y = 5 (11) + 70 = 55 +70 = 125.

Extrapolation would be if we used the equation to predict a value below x = 6 or above x = 14. Such results would be unreliable because we don't know for sure what the results might be outside the data range ; they may not continue as a line.

NOW YOU TRY 7 :

i) Plot the points below,
ii) Draw a line of best fit.
iii) From the line of best fit derive an equation
iv) Use your equation to interpolate a result for x = 3

x	0	1	4	7
y	0	3	8	11

Trends and Variation

Trends can be linear (straight lines) or non-linear . An example of a non linear trend might be a curve . In the diagram below we see a non–linear trend that shows acceleration.

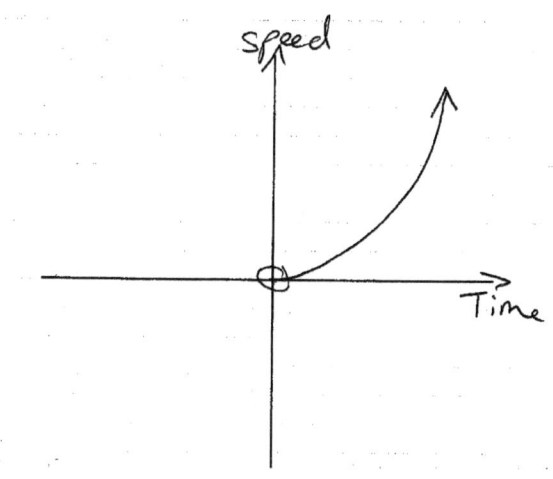

Direct vs. Partial Variation
When there is a pure multiplicative relationship between quantities, we call this direct. On a graph this is simply any line that starts at (0,0) such as the one below.

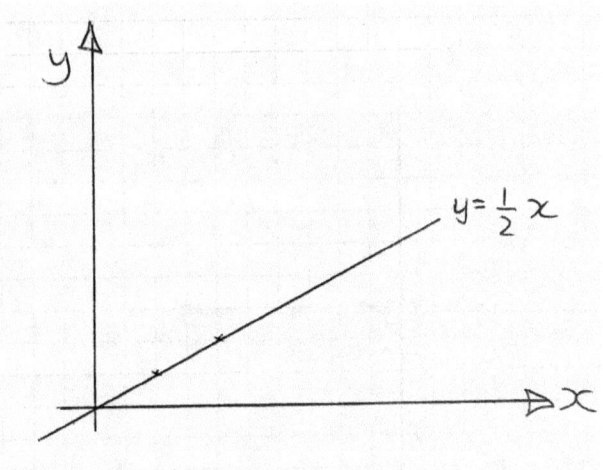

Partial Variation is a linear relationship, but one that starts above (or below) 0,0.

 In the example below the line has the same gradient but the y-intercept is 1 instead of 0, and we call this starting point <u>the initial value.</u>

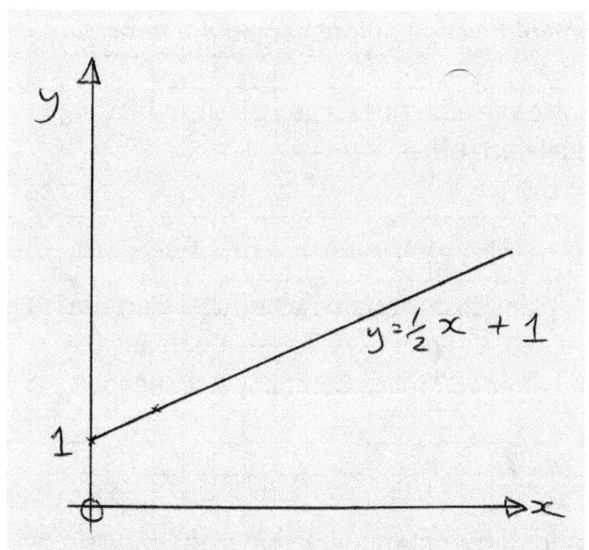

Real life applications of Partial and Direct Variation

A classic example of Partial vs. Direct variation might be differing cell phone plans.

Company A might offer a monthly deal whereby a set charge of 30 dollars was applied every month, and every minute was 0.30 dollars.

Company B might offer a deal whereby there was no set charge but each minute costs 0.50 dollars.

Company A could be represented by the partial equation y = 30 + 0.50x

Company B could be represented by the direct variation equation y = 0.50x

If these equations were graphed, their intersection point

would be when the companies charged the same amount.
We could solve the equations algebraically by setting them to equal each other.
$30 + 0.30x = 0.50x$
$30 = 0.20x$
$x = 30/0.20$
$x = 150$. The plans cost the same when 150 mins are used. Before 150, the graphs will show Company B is cheaper, and after 150 mins, Company A is cheaper.

NOW YOU TRY 8:

Two gyms offer different membership plans. 'Muscle Gym' offers a flat fee of 20 dollars per month and charges 2 dollars per visit. 'GO Gym' charges no set fee but costs 5 dollars per visit.

   i) Plot both options on the same graph and determine at how many visits per month the gyms would cost a member equally.

   ii) Explain how the graph would look if Muscle Gym dropped it's flat fee but raised it's per visit charge to 4 dollars.

Linear equations have variables that are only to the power of 1. Of the two equations below, (i) is linear and (ii) is non-linear :

i) Y = x
ii) Y = x²

Linear Equations (equations involving 'x' to the power of 1 only) can be plotted on a graph as straight lines. A table of values is useful to record x values (horizontal moves on the graph) and the corresponding y values that pair with them (vertical movement on the graph)

For example, to plot the equation y = 3x – 4 on a graph :

Pick an 'x' value (any will do, but you want something near the origin (0,0) so x = 0 is a good start.

Substitute your x value into the equation.

If y = 3x – 4, then if x = 0

y = 3 (0) - 4 = - 4.

x= 0, y = - 4 can be written as an ordered pair', or co-ordinate on a graph. (from the middle, do not move left or right as x = 0, but as y = - 4, go up – 4.)

Repeat the procedure with another x value, eg. Pick x = 1 and sub into y = 3x – 4
If x = 1 , y = 3 (1) – 4 = -1
So we have the point (1, -1) to plot.

Now that you have two points you can join them up and

extend a line. This is the line of the equation.

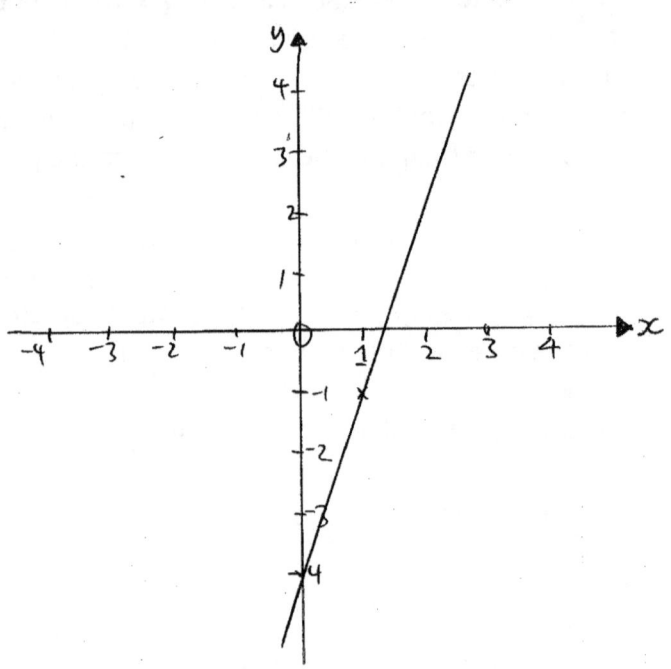

NOW YOU TRY 9 i) *Plot the equation $3x - y = 5$ as a line on a graph*

Y = mx + b : the general equation of a line

When plotting equations, you can find the gradient or **slope of a line** by looking at the number in front of x *if the equation is in the form $y = mx + b$.*

In the above example, if y = 3x - 4, then the slope is +3. For every one unit you go across, you go up 3 units.

If an equation is not in the form y = mx + b then you can rearrange it so it is.
E.g if given   2y – 4x = 6, rearrange for y :

2y = 6 + 4x
y = 3 + 2x . Now you can tell that the slope is 2 (the number if front of x).

The general equation of a line also tells you the y-intercept ; the point at which the line crosses the vertical axis. In the equation this is the constant ; that is the number on it's own
In the equation y = 3x – 4, the y-intercept is - 4.

NOW YOU TRY 10:
Find the slope and y-intercept : i) 2y + 7 = x   ii) 1/2 y - x = 4

Application of slope and y- intercept to Real Life situations :

Since slope is a measure of rate of change, in real life situations it can be used to model a variable such as unit cost.
The y –intercept is used to model a constant such as initial value or fixed cost (you saw such examples with

direct and partial variation).
For example, a cell phone plan costs 30 dollars a month plus 0.10 dollars per minute.
The 30 would be the y-intercept and the 0.10 would be the slope as each minute adds a unit cost of 0.10
A pricing plan could be modelled with $y = 0.10x + 30$
If the unit cost was raised the slope would get steeper.

Finding intersections of lines using a graphical method

a) <u>Graphical Method</u> :

Plot each equation by using a table of values or other

e.g : Solve i) $y = 3x + 1$, and ii) $y = 2x - 4$.

For $y = 3x + 1$, try $x = 0$;

- If $x = 0$, $y = 3(0) + 1 = 1$. So we have the point (0, 1).

Repeat for another x-value of $y = 3x + 1$:

- If $x = 1$, $y = 3(1) + 1 = 4$. Hence we have the point (1, 4).

Plot both these points and join them up to draw the line of $y = 3x + 1$.

It should look like the diagram below.

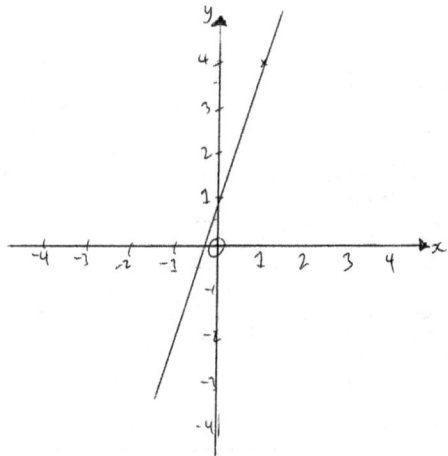

For y = 2x − 4, first try x = 0 :

If x = 0 , y = 2(0) - 4 = -4. So we have the point (0, - 4).

If x = 1, y = 2 (1) − 4 = -2. So we have the second point (1, -2)

- Plot this line on the same graph :

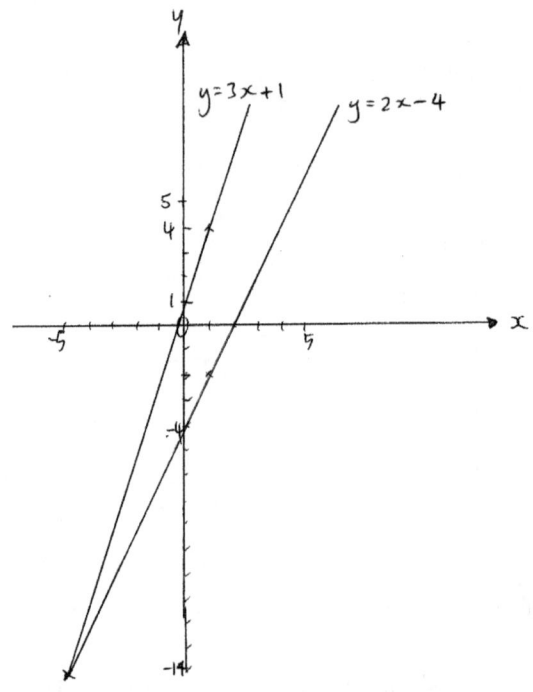

The solution is where the 2 lines intersect ; in this case (-5, -14).

At this particular point both equations have the same x and y values.

Real life applications of intersections of lines

Intersections of lines are usually used to model competing pricing plans and see at what usage point they are equal.
For example, two internet providers offer different monthly packages ;
Linkage offers a flat fee of 20 dollars plus 0.30 dollars per Gb downloaded, whilst Kinekted offers a deal of 40 dollars per month with 0.10 dollars per Gb downloaded.
Linkage could be modelled using C = 20 + 0.30g
Kinekted could be modelled with C = 40 + 0.10g
The usage point at which both plans are equal cost can be found by graphing the lines and finding their intersection point.

NOW YOU TRY 11 :
Complete the above problem and find at what usage point the plans cost equally, and what this cost amount is.

Section C : Measurement and Geometry

The Pythagorean Theorem

i) In a right triangle, the two short sides squared and then added equal the longest side squared. You can then square root this sum to find the long side.

eg. - If a triangle has short sides 8cm and 6cm, then the long side (the hypotenuse) can be found using $8^2 + 6^2$ = 64 +36 (make sure you square the numbers *before* you add them ; remember BEDMAS !)

$\quad$ 64 + 36 = 100
$\quad \sqrt{100}$ = 10 ; the long side is 10cm.

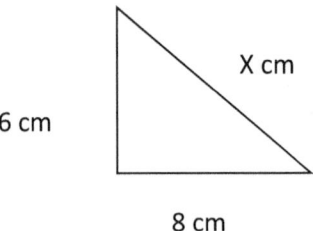

6 cm

8 cm

ii) $\quad$ You can re-arrange the formula $a^2 + b^2 = c^2$ *to find short sides*.

The formula can be rewritten as $c^2 - a^2 = b^2$, or long side squared *minus* short side squared gives the other short side squared.

e.g : A triangle has hypotenuse 15 cm and width 9cm. The missing side can be found with $15^2 - 9^2$ = 225 – 81
225 – 81 = 169
$\sqrt{169}$ = 13 cm

NOW YOU TRY 12 :
i) A right triangle has width and height 3m and 6m respectively. Find the length of the hypotenuse
ii)   A right triangle has hypotenuse 8cm and height 7cm. Find the width.
iii)   A right triangle has hypotenuse 11m and width 9m. Find the height.

Surface Area and Volume

Most of these problems involve 3-D shapes. Think of volume as how many slices you have of a shape, and surface area as the area of the outside faces only. You will often have to apply trigonometry to these problems.

Eg. 1 – Surface area and volume of a simple rectangular prism that measures 3cm (height) by 4cm (length) by 5cm (width).

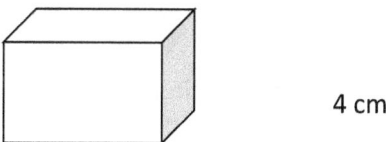

4 cm

c

The surface area is the total of the faces.
Note there are two of each face, e.g front and back are the same, the sides are the same, and the top and bottom are equal too.

S.A = 2 [(l x w) + (l x h) + (w x h)]
S.A = 2 [(4 x 5) + (4 x 3) + (5 x 3)]
S.A = 2 [20 + 12 + 15] = 2 x 47
S.A = 94cm².

The volume can be found simply using V = l x w x h
or V = 5 x 4 x 3 = 60cm³.
It is also useful to think of taking a 'slice' of the shape and multiplying by the 3rd dimension. The area of the front face is 3 x 5 = 15cm². Now multiply by 4cm, or imagine 4 slices through the shape.
V = 4 x 15cm = 60cm³.

Cylinders involve a slightly different formula but the idea of multiplying by slices is the same. Look at a cylinder of diameter 8 cm and height 6cm.

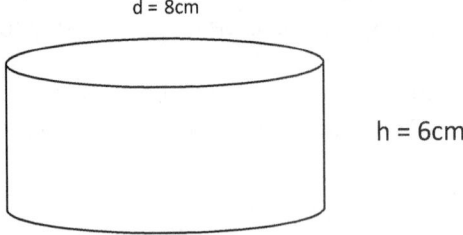

Find the area of the circle top or bottom using pi x r².
In this case A = 3.142 x 4² = 157.95 cm²
Multiply this answer by the height 6 cm to find volume :
V = 6 x 157.95 = 947.73 cm³ (Note * : I have rounded answers to 2 d.p to save space, but you should only round FINAL answers. Use the 'Ans' button on your calculator to use previous answers and retain accuracy.)

To find surface area of a cylinder use the circumference times the height to find the 'wrap around' section (if you peel a label off a can you will see this is height x length where length = circumference).
For the case above the curved surface area is 6cm x

3.142 x 8 = 150.816 cm²

<u>Volume of cones</u> :, Cones are just one third of a regular prism that has the same height and width.
i.e A cone has one third the volume of a cylinder that has the same dimensions. So just use the cylinder formula then divide by 3.

<u>Surface area of a cone</u> *uses the slant length instead of height,* and you may need to use Pythagoras with the height to calculate slant length.
Imagine a Right Triangle slice sitting inside the cone. Slant length is then used to calculate the curved part of the cone using
Curved area = pi x r x S.
 In this cone the slant length can be found using the radius as the base.
$S^2 = h^2 + r^2$, in this case $S^2 = 4^2 + 2.5^2$ = 4.72 cm

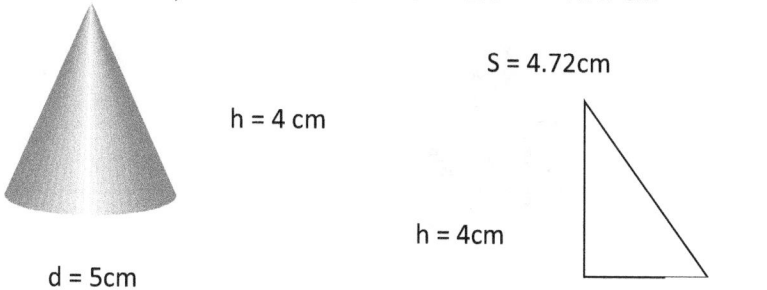

S = 4.72cm

h = 4 cm

h = 4cm

d = 5cm

R = 2.5 cm

Slant length (S) is now used in the surface area formula
: S.A = pi x r² (for the circle base)   + (pi x r x S).
S.A = (3.142 x 2.5²) + (3.142 x 2.5 x 4.72) = 56.71 cm₂

A pyramid uses the same principle as a cone for volume ; just find one third the volume of a regular prism that has the same dimensions.

<u>The surface area of a pyramid</u> involves finding the slant length which is used as the 'height' of a face. *In this case slant length means the line which runs down the middle of each face.*
Like a cone, use Pythagoras with half the width to find slant length.
In this case use height 5cm with width 6cm : $S^2 = 5^2 + 3^2$; $S = \sqrt{34} = 5.83$cm
Once you found the slant length use it as the height of each face.
In this case the area of each face is $A = \sqrt{34} \times 6 = 35$cm.
The total area of a square based pyramid is the base plus 4 times the face area.
S. A = $(6 \times 6) + 4 (\sqrt{34} \times 6) = 175.94$ cm²

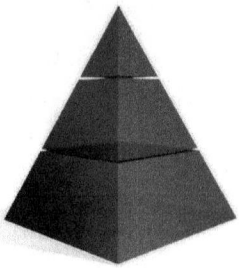

6cm

For balls or Spheres use S.A = $4 \times 3.142 \times r^2$ (think of 4 circles)
For volume of spheres use V = $4/3 \times 3.142 \times r^3$
(remember the 3's are in the volume formula)

Sample Mixed problem : A cone has a height of 16 cm

and diameter 10 cm. What is the surface area in square inches ?

- We need slant length, which is $S^2 = 16^2 + 5^2$; S = 16.76 cm
  S.A in cm is : S.A = (3.142 x 5^2) + (3.142 x 5 x 16.76) = 341.90cm². Since 2.5 cm = 1 inch, 341.90/2.5 = approx. 137 inches

   ### *NOW YOU TRY 13:*
   I) *A swimming pool measures 8 feet by 4 feet by 20 feet. What is the volume of water it can hold in litres ?*
   ii) *A basketball has a diameter of 28cm. What is it's surface area in cm, and how much air can it hold in litres ?*
   iii) *A cardboard pyramid has a square base of 5cm by 5cm. Each face has a slant length of 7cm. Find the volume of the pyramid.*

Maximisation of Area given Perimeter

In general, *the more regular and symmetrical a shape is, then the more efficient it is* at either maximising area or minimising volume. Consider for example the simple case of how to arrange 20 metres of fencing so as to enclose the biggest area (Imagine for instance you wanted the biggest possible space for a farm animal).

Look at some possible arrangements of Length vs. Width :

$1 \times 9 = 9 m^2$
$2 \times 8 = 16 m^2$
$3 \times 7 = 21 m^2$
$4 \times 6 = 24 m^2$
$5 \times 5 = 25 m^2$
$6 \times 4 = 24 m^2$

You can see that using 5m x 5m gives the biggest area. In fact, for any four sided shape, the square will always give the biggest possible area, and for any shape, having the same width as length will maximise area. The 'perfect' shape is the circle, as it is the same shape in every direction

NOW YOU TRY 14 :
60m of fencing is to be arrange against a wall to form a rectangular or square shape.
 i)How long should each side be to maximise enclosed area ?
ii) What will be the resulting maximised area ?

Minimisation of Area given Volume

Symmetrical shapes are also the most 'efficient' in 3 dimensions. A cylinder for example that has the same radius as height will contain the maximum volume given a certain limited surface area.

A cube for example that has equal length, height and width will hold the maximum volume compared to total surface area. Imagine we need to build a box that holds 200cm³ using the minimum amount of cardboard. If width = length = height, then we could model the problem with :

$(x)(x)(x) = 200cm^3$
i.e $\quad x^3 = 200$
$\quad\quad x = \sqrt[3]{200}$
$\quad\quad x = 5.85$ cm (2 d.p)

NOW YOU TRY 15 :

A tin manufacturer wants to produce cans that use the minimum amount of metal yet hold the maximum volume. Given that the tins are to hold 500ml , what should the dimensions of the radius and height be ?

Angles in Polygons

A polygon is a straight sided shape, such as triangle, but not a circle.

Any polygon can be broken into triangles, and this helps you calculate the total interior degrees, as each triangle has 180° inside.

A polygon always has 2 less triangles inside than it's number of sides. For example, a five sided shape can be split into three triangles. Therefore, a pentagon has 3 x 180 degrees inside.

Total interior degrees of a pentagon = 540°.

The exterior angle of a polygon is 180° minus the interior.
In a regular pentagon, there are a total of 540° and five sides, so each interior angle is 108°.
Each related exterior angle is 180°-108° = 72°

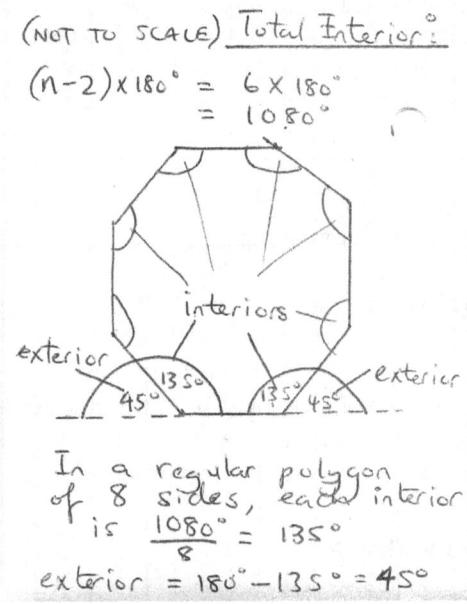

Exterior angles are also directly related to the number of sides of the polygon.
360° / exterior angle = the number of sides.
e.g
In a hexagon each exterior angle is 360° / 6 = 60°
Note how you could also then quickly find the interior 120°.

NOW YOU TRY 16 :
Calculate the i) exterior and ii) total interior angles in a

Nonagon (9 sided shape).

Angles around a vertex or corner can generally be calculated by using 4 fundamental geometry rules :

    i)   The angles on a straight line add to 180°
    ii)   Angles opposite crossed lines are equal.

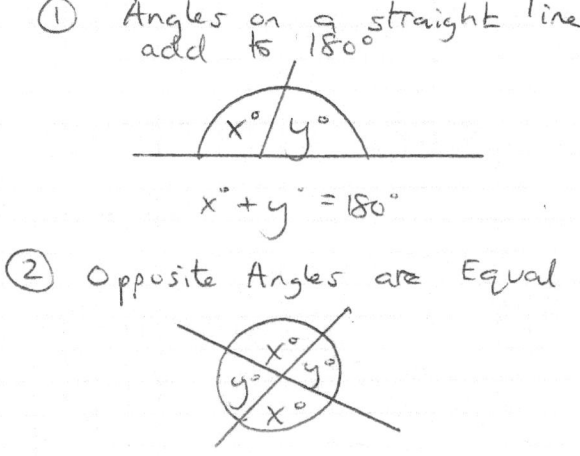

To these rules are added 2 more useful Angle Geometry principles that occur *when parallel lines are crossed or traversed by the same line* :

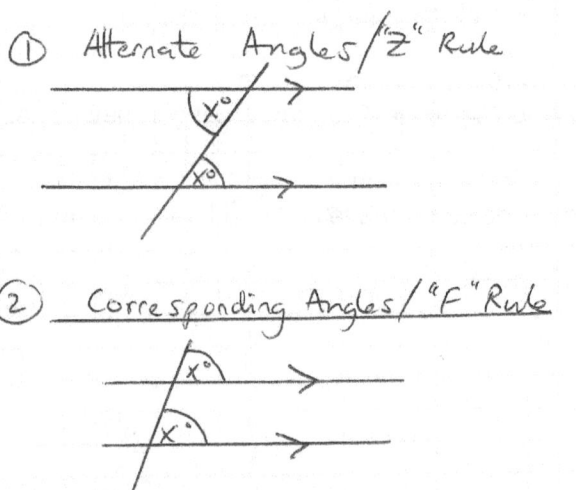

① Alternate Angles / "Z" Rule

② Corresponding Angles / "F" Rule

NOW YOU TRY 17 :
Apply the four angle rules above to find the labelled angles in the diagram below :

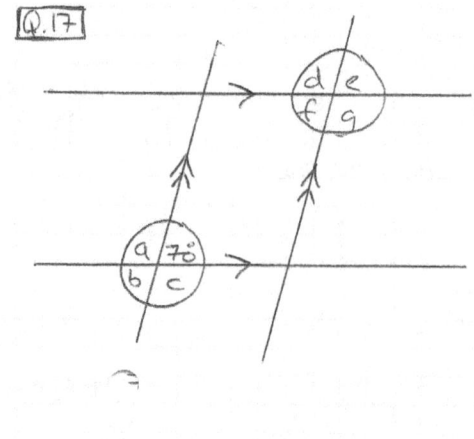

Answer Sheet :

CHECK YOUR ANSWER 1 : i) 25/4   ii) 54.76

CHECK YOUR ANSWERS 2 :   i) $V = 6x^3$, S.A = $22x^2$ ii) $48x$ power 5   iii) $5x^3$ iv) $5x^2y$

CHECK YOUR ANSWERS 3: i) -2   ii) 6.5 %   iii) 200g : 400g : 600g

CHECK YOUR ANSWERS 4:   x = 2.64

CHECK YOUR ANSWERS 5 : i) $r = C/2\pi$ ii) $r = \sqrt{(A/\pi)}$

CHECK YOUR ANSWERS 6 : i) x = 24   ii) x = -2.5   iii) x = 2.25   iv) x = 1.63

CHECK YOUR ANSWERS 7 : i) y = 5   ii) y = 1.5 x + 1

CHECK YOUR ANSWER 8 : i) 6.67 hours   ii) it would be a direct variation graph but steeper.

CHECK YOUR ANSWER 9:

CHECK YOUR ANSWER 10 : i) m = 1/2 , b = -3.5   ii) b = 8, m = 2

CHECK YOUR ANSWER 11 :   C = $50 , g = 10

CHECK YOUR ANSWER 12 : i) 2.6cm   ii) $\sqrt{15}$cm   iii) 2.51cm

CHECK YOUR ANSWER 13 : i)
8 x 4 x 20 = 640ft³. Using 1 foot = 12 inches = 30.48cm , 640 x 30.48³ = 18122781.82cm³ = 18122781.82ml = 18122.78182 L
ii) V = 4/3 x 3.14 x 14³cm

= 11488.21333cm³ = 11488.21333ml = 11.49L
S.A = 4 x 3.14 x 14² cm = 2461.76cm²
iii) V = area of sq. Base x height.
   Using pythagoras, h = 2.56cm
   V = 25 x 2.56 = 64cm³

CHECK YOUR ANSWER 14 :
width = length ;   x + x + x = 60 ; x = 20.
ii) Area = 400m²
CHECK YOUR ANSWER 15 : maximum volume will be diameter = height. Use h = 2r or r = 0.5h
    3.14   x r² x h = 3.14 x 0.5h² x h = 1.57h³
Use 500 = 1.57h³
h³ = 500/1.57
h = 6,83 (2 d.p)

CHECK YOUR ANSWERS 16 : i) 40° ii) 140°
CHECK YOUR ANSWERS 17 :
a = 110°
b = 70°
c = 110
d = 110
e = 70
f = 70
g = 110°

www.ingramcontent.com/pod-product-compliance
Lightning Source LLC
Chambersburg PA
CBHW061520180526
45171CB00001B/270